VEDIC MATH

John Carlin

Copyright © 2014 John Carlin

All rights reserved.

ISBN:1501075187
ISBN-13:978-1501075186

CONTENTS

1	**Vertically and Cross-wise**	**1**
2	**By One More**	**23**
3	**Bi-directionality**	**30**
4	**Over/Under 100**	**37**
5	**Over/Under 20,30...**	**42**
6	**More on Ratios**	**51**

1. Vertically and Cross-wise

Let's review the general method that we use to do mental math. If you have taken algebra , you already are acquainted with the form that is used. Multiply your two "factors" as though they were in algebraic form. In Vedic math this is called "Vertically and Cross-wise". It is in fact one of the 16 Sutras of Vedic Math. This method allows you work from left to right. First, do the vertical multiplication of the first two digits, then cross multiply and add, finally, do the vertical multiplication of the last two digits.

If you were multiplying 23x12, you would start out by multiplying 2x1, then (2x2)+(3x1), finally, multiply 3x2. The components to your multiplication would be 02/07/06. Your answer would be 276. Using this method you can multiply any algebraic expression. What is particularly exciting about this method is that you can write your answer down from left to right on a single line.

This fact alone gives you a huge advantage when multiplying compared to the grade school method you learned. You can assemble the parts in your head as you go. The method also allows you the opportunity to look at the relationship of the parts to each other. The nature of the relationships is a huge part of mental math.

Those relationships allow you to do shortcuts and give you a math agility most people confined to the grade school method never develop. Indeed, I almost never use the grade school method you learned to do math. I do use a pencil and paper, but always think in terms of components and relationships. Very simply, you can't beat a system that involves the use of your mind as opposed to the mindless use of single method such as the grade school method.

With two and three digit multiplication you will always have three components to assemble. If you put them together in order from left to right , you are basically writing the answer down as you calculate. In the case of three digit multiplication you split the three digits into a two digit number and a one digit number. Generally , you would split both numbers the same way. For example , if you want to multiply 112x103, you could split the 112 into 01 and 12. Then split the 103 into 01 and 03. Now proceed to do your vertical and crosswise calculations. The components would become 01/15/36. Your answer would be 11536.

Conversely you could have split 112 into 11 and 02. You could have split 103 into 10/03. In this instance your components would be 110/53/06= 11536. Three digits is no more difficult than two digits, you have the opportunity to split the number in a fashion that is easiest for you. The only thing that is different is that you will have two one digit by two digit multiplications to do and add together to get your cross product.

In that sometimes that cross product calculation is not easy, we have a myriad of ways to make it simpler. Finding the cross product is the hardest part of mental math. For one thing when the number one is involved we have another Sutra we can apply. That one is called By One More Than The One Before. This book is largely about the application of that sutra. The method works to multiply numbers close to a base of 1, 10, 100, or more, you note the incremental amount it is over or under a base of one, ten, or more. Then you work off of that. In the case of the problem we just solved we could say one number was 12 units greater than 100, and the other was 3 units greater than 100. Now we either add 3 to 112, or add 12 to 103. Either way we get 115. Now we just tack on the product of 12*3=36. Our answer is 11536.

We will be going into this in much more detail in the ensuing chapters, it is quite simple, and effective for situations where you are close to a base especially. View it as a different tool that will come in handy when you want to do a quick calculation.

Being close to a base was one of the topics I didn't get to in my first book *Using Fractions to Multiply*. One of the reasons I didn't cover that topic was that I had already thrown so much out there in regards to math. I wanted to confine that book largely to a single topic. That topic was fractions, ratios and the use of those to help you become a better mental math calculator.

Along the way we covered far more than that. This was largely because you can't talk about ratios and fractions without touching upon a whole bunch of other things. In this book we are going to see how the identity property is way underrated as to it's power and application to mental math. The identity property of multiplication means that you can multiply any real number by the number one and not change its value.

You would say, and I would have to concede that this definition seems lame and worthless. By the time we get done working on multiplication close to a base of one, ten, or one hundred you will appreciate how useful this simple property is.

Along the way we are going to touch on a lot of other things again. Hopefully, I can touch upon some of the same items we talked about in the first book in a way that briefly covers those items , and yet elaborates on them as well. Just as we did in the other book we will start off by discussing single digit by double digit multiplication, this time from the perspective of the identity property.

Then we will go from there to see how it is bidirectional and symmetrical. We will see how the technique we're going to learn can be combined with the use of ratios. For sure we will be elaborating even more on the use of digit sums to multiply, and synchronized ratios. We will also look at a couple of my favorite numbers nine and 11, and we will elaborate on their properties even further.

One of the key skills you need for mental math is the ability to multiply a double digit number by a single digit. There are all kinds of shortcuts that you can use to help you do this in your head. Multiplying a double digit by a single digit involves linking two multiplications together, one offset from the other by a multiple of 10.

For example, if you wanted to multiply 9x31. You could easily do that in your head by multiplying 9x3= 27. Now you link that 27 with the product of your second calculation 9x1, which equals nine. The numbers you are linking then are 27/09. Clearly, you can see that the zero was important to help you place keep. Our answer is 279. Note the identity property came into play on this one, in the form of 9x1= 09.

Since ratios and proportions are so important to mental math, my suggested way to approach this problem is not to just grind out the numbers. Just take a second and think of 31 as a fraction. That fraction would be 3/1 or the whole number 3. if you look at it from the other direction . The fraction is one third. What I'm getting at is that if you know one of the two multiplications and the relationship of the numbers to each other. You don't have to do the second multiplication.

In this case, the lead numbers 27, in the second number is 27 divided by three, or nine. Tack the nine onto the 27, and you have your answer 279. This is a subtle departure from what you were doing previously. But the distinction is very important. What you're really doing is recognizing a pattern or relationship first, and then working off of that. Considering your double digit number as a fraction is the starting point for using fractions to multiply.

When the relationships are real simple this works pretty easily. Anything that begins or ends with one makes a whole number and it's reciprocal is a easy fraction. The math is about as simple as you can get.

Later on we will find out that double digit multiplication that begins or ends in one is just as easy. Relationships where one number is twice the other, or three times, are also quite simple.

For example 48x9. Nine times four equals 36. The second number is merely two times that amount. So you're linking 36/72 to get 432.

Nine as a multiplier is particularly interesting. Perhaps you've heard that casting out nines is a good way to check your answer in multiplication. It's actually even more useful than that. Anything you multiply by nine will have a digit sum of nine. The digits in 432 add up to nine. The components add up to nine too. Nine combined with the number 11 also has some interesting characteristics.

If you are in middle school you have taken, or are currently taking algebra, or will take algebra. When you run into algebraic equations where (x +1) is a factor, it will always turn out that the sum of the odd coefficients will equal the middle coefficient. The mental math equivalent of this would be multiplying a number by 11. Anything you multiply by 11 will have middle coefficients , or components that add up to the sum of the odd coefficients. 11X23.

For example is 253. The two and the three added together equals that middle number five. Sometimes when the numbers are larger, this fact is hidden because we go over 10 in the middle and carry one of the digits over to the lead number. It is also true that (x-1), which is the algebraic equivalent of nine has the characteristic of always giving you an answer that adds up to nine.

In algebra, this manifests itself in a distinct way. The odd coefficients added together with the middle coefficient always equals zero. The algebraic equivalent of multiplying 48 x 9, would be multiplying the factors of the two equations (4x +8) * (x -1). The expansion of this problem, would give you $(4x^2+4x-8)$ notice how the sum of all the coefficients equals zero?

The other easy tip I wanted to mention in regards to nine is about multiplying a two digit number by it that has a digit sum of eleven.

For example 9x92, or 9x83. The numbers that you are linking together are reciprocals of each other. There is no need to do the math so to speak beyond the first multiplication. In the first case you are linking 81/18, in the second case you are linking 72/27. The two outside "wings" are equal, the inside "body" is twice the last digit. So you have an answers of 828, and 747.

This is not unlike multiplying two numbers together that are reciprocals. 72*27 would have components of 14/53/14. Again the outside wings are identical, they are the product of 7x2. The middle body part is always the sum of the squares of 2 and 7. In algebra whenever you see first and last coefficients that are identical in value and sign you should look for a possible reciprocal relationship.

From an algebraic standpoint 27x72 is the same as $(2x+7)*(7x+2)$, where x=10. The expansion would be $14x^2+53x+14$. If you were factoring this expression you could rule out a lot of things very quickly. The factors of 14 and 1 are not worth considering. The factors of 7 and 2 are the ones to consider 7-2=5, 5-2=3. There you have it by observation.

From the factors you also get the roots. The whole point being you want to contemplate a relationship, and not get bogged down grinding out the factors. From a fraction standpoint you are looking at 7/2+2/7 which equals 49+4/14. Everything you need to get your answer is contained in that expression. The components 14/53/14= 1944.

There is a whole class of problems, namely two digit squares where fractions are very important. For example, if you wanted to multiply 12x12. You could do it by considering each number as a fraction. In effect, you would be adding 1/2+1/2 to get a whole number of one. That means in this case that there is a one to one relationship of the product of the denominators to the middle component of our calculation. We know the 2x2 equals four, just knowing that and that one times one equals one leads us to know that 12x12 involves linking together the following three components 01/04/04 =144.

Normally to get that second component, or middle

coefficient you would have to cross multiplying and add (1*2)+(1*2)= 04. You can skip all that laborious work by just knowing that you have a one-to-one relationship with the product, 2x2. If the problem were turned around in you are looking to find the square of 21. You could again look 21 as a fraction. 2/1+2/1= 4.

This time the middle coefficient is exactly equal to the product of 2x2, the lead number product. We did one multiplication and got to the three components we need to do the math in our head. In this case we will end up linking together 04/04/01= 441.

Vedic math is partially an algebra-based math that allows you to mentally do significant calculations in your head. Many of the rules for manipulating the numbers are based on little rhymes are sutras. There are 16 of them altogether., and there are some sub sutras as well. I urge you to google vedic math, visit some of the web sites out there on the subject. It is fascinating how easy this method makes math.

You are not limited to fractions that add up to one to use fractions to multiply. The can add up to other whole numbers. 42X42. For example, would give you the whole 8/2+8/2= 4. Or looking at the fraction the other way would give you ¼. So immediately you know that you are linking 16/16/04 to mentally get 1764.

The use of fractions can be extended to include numbers that are not squares. For example, 13x23. In fraction form you are adding 1/3+2/3 equals one. That means your components are 02/09/09= 299. The use of fraction form does not limit you only to the whole number one. You could have other whole numbers, you can have whole numbers plus a fraction and still quickly come up with components. It really pays to look at the numbers as a fraction before you dive in and start trying to arrive at the answer.

We can take the 13x23 example and extend the use of fractions to include multiples. For example 26 is still the fraction 1/3. 26x23 would still give you two fractions that add up to the whole number one. In this case you don't really have to add the fractions, just recognize that they add to one so now you know that your middle component is still the same as the product of the denominators 6x3. The components and answer are 4/18/18= 598.

In the case of three digit multiplication you just look at the three digit number as an improper fraction, or a fraction with a two digit denominator. For example, if you wanted to multiply 412x812. You could look at the sum of the fractions (8/12)+(4/12). Since they add up to one you know without doing any math that the middle component is going to be the same as the product of the denominators multiplied together. So now you just have to put together 32/144/144= 334,544. The number 612 you could square in your head to get 374544. you just did real three digit multiplication in your head, and it was real easy!

So there you have it a quick review. Nine and eleven pass genetic traits onto, or into everything they multiply with. Nine has a secret reciprocal relationship with all the two digit numbers that have a digit sum of 11 that it multiplies with. I might add that nine has a lot of secret relationships beyond that. You can predict the second product of any two digit number you multiply by nine by just looking at the digit sums. Reciprocals have a special pattern as well. Beyond that there are the older shortcuts everyone knows about.

You can square any two digit number that ends in five by multiplying the first digit by one more than it's value, and tacking 25 onto it. 25X25 would be 3*2, with 25 tacked onto it to get 625. 35x35 would be 4x3 with 25 tacked onto it to get 1225. This same "by one more" sutra can be used when you multiply any two digit numbers together that have the same tens base and units that add to 10. For example, 81x89 would be 9x8 plus tack on 09 to get 7209.

The same units and base approach can be extended to include some very powerful methods. Notice how "by one more" comes into play on this example. If you wanted to multiply 21x39 you could note that the units digits add up to ten and the larger tens digit is one more unit greater than the small digit. In this case, your cross product needs no calculation it is in fact right under your nose. It is in fact the smaller number itself.

Anytime you have units that add to ten and tens units that are one unit apart the smaller number itself is the cross product. In the 39x21 example skip doing any fractions or cross multiplication to get the cross product. In this case the components are 6/21/09= 819. If you were multiplying 82x98 you can virtually write your answer down from left to right as 72/82/16= 8036.

Notice that we can also look at 39x21 and borrow a set of ten from the larger number. By doing that we make the 39 into 2/19. Now the two numbers have the same base of 2 but the units digits add up to 20 now:

```
   2  19
   2  01
  08/ 19= 819
```

In this case, we multiplied our base by 4 more and tacked on the 19.

If the question had been to multiply 49x21 we could do a similar procedure and turn 49 into 2/29. Now we have 2/29 and 2/01. The base is the same and the units add up to 30. We multiply the base by the base of 20+30=50. So our new lead number is 10 with 29 tacked onto it. We have expanded this little shortcut to be very useful even when the initial precondition of having the same base is not immediately apparent.

In the three digit world the same principle works also. If you wanted to multiply 248x252 you could note that 48 and 52 add to 100. By one more would mean you multiply the 2 by 3 to get 6. Now tack 48x52 onto that. Multiplying this in your head is pretty easy, the two numbers add to 100 and average out to be 50. Each number is two units from 50. 50x50=2500. Now subtract adjust the 2500 by subtract 2x2 from 2500 to get 2496. Tack 2496 onto 6 to get 62,496 as your answer.

If the two numbers being multiplied were 219x211 we could also multiply 21x22 to get 462 and tack 09 onto that to get 46,209 as our answer. Really all that happened by going to three digits is that we expanded the opportunities we have to apply our shortcuts to finding a solution. It's the old story of difficulty creating opportunity.

        The bi-directionality principal applies too. We will be reading about that in the next chapter. Basically it means that what works in one direction works in the other. If you were asked to multiply 39x21 backwards you would be multiplying 93x12. You could note that the tens units add up to ten and the units are one unit apart. So now your calculation of components would be 09/21/06= 1116.

        My cross product is still 21, I just had to read it backwards or from right to left. It's still there though. This one is so powerful it is incredible. We doubled it's power with the fact of bi-directionality, and we can extend it to three digit numbers also. Consider for a moment what happens when the total of the units column is not ten. Very easy to adjust for. Each unit over ten or under ten is 1/10th. We will definitely be looking at this in a chapter all by itself. So without further rambling let's look at more math tools and opportunities.

2. By One More

In actuality, you never really needed to learn the multiplication tables past 5x10. Let's say you wanted to multiply 9x8 but didn't know your tables that far. You could do the multiplication anyway, by taking the compliment of these two numbers. That would be one and two, respectively. Since these two numbers are compliments, they represent the deficit from 10 that we have. The word deficit implies that they're both negative numbers in regards to the base 10.

If you want you could represent these two numbers on your fingers. Put one and first finger of your left hand, and two on the first two fingers of your right hand. In any event, we can now subtract two from nine, or one from eight. Either way, we get seven. Hold that seven in your mind for a second, while you multiply the two numbers on your fingers together. Those two numbers are -1 and -2. Multiplied together they equal +2. Let's tack that two onto the seven to get 72. 72 is of course the product of multiplying nine times eight. What just happened? We actually multiplied 7x8 without using the times tables beyond multiplying one times two.

Let's do the same thing with 7x6. Put -3 on your left fingers. Put -4 on your right fingers. Now subtract -4 from seven, or -3 from six. Either way, we get three. Multiply the numbers on your fingers together to get 12. Now put 3/12 together as though three were 30, and we were adding 12 to it. And give us 42. 42 is in fact 7x6.

All this is well and good, but you already know times tables past five. So what value is it? Well, this is the starting point for learning to use the identity property to help you multiply. In the background there is the fact that we were working with the number 10 as a base. The numbers 1, 10 , 100, 1000, and indeed 10,000 and beyond are all subject to the identity property. Let's leverage this concept a little more.

If you wanted to multiply 6x14, you could do it in a similar fashion to what we just did. Note that six is four under the base of 10. 14 is four over the base 10. We have a common base, let's put four on the fingers of one hand, and four on the fingers of the other hand. Now we can either add four to six, or subtract four from 14. Either way you get 10. Now multiply those numbers on your fingers together. 4X4= 16. since one number is over a base in the other is under a base. We are dealing with a negative number, times a positive number. That resultant 16 is a negative number in this case. Subtract 16 from 100 and you have 84 , which is in fact 6x14.

Let's do another example. This time, let's try 8x17. One number is two under a base of 10. The other is seven over a base of 10. Put the number two on your left fingers. Put number seven on your right fingers by using your thumb as of five +2 more fingers. Now either subtract two from 17, or add 7+8. Either way you get 15. Hold that 15 in your head for a second and multiply the two numbers on your fingers together 2*7= 14. So now we have 15/-14 to deal with. That would be the equivalent of 150-14= 136.

Now let's try an example where both numbers are over the base of 10. Let's multiply 13x19. Put three on the fingers of your left hand, and nine on the fingers of the right hand. We are using a Chinese Abacus method with our fingers for numbers over five. In this case your thumb is the five bead and all your remaining fingers extended represent nine. Since everything is over a base of 10 everything is additive, or positive. We are adding 9 to 13, or adding 3 to 19. Either way , we get 22. Now we link the 22 with the product of our fingers. That would be 3*9=27. So we are linking 22/27 to get 247. All this can be done mentally with a little help from our fingers. Is this pretty cool or what?

Let's do one more where both numbers are over a base of 10 just to make sure we have this method down. Let's do (19*19). This should be pretty easy. Both numbers are nine over a base of 10. Add 9 to 19 to get 28. That is the first half of the answer. Now we link it with the product of the nines we have on the fingers of each hand. 9*9 equals 81. 81, then is the second half of our answer. We have to link 28/81= 361. We just squared 19 mentally.

I don't know if the Karen Carpenter song *We've Only Just Begun* is still a standard at weddings these days, but as the song says we have only just begun. Soon you and math will be together in bliss! Firstly, let's do some exercises using the techniques we just learned. You can check your answers with a calculator if you like. I would prefer that you use fractions, and some of the other methods we learned in my previous book to check your work. Below is a example template for that.

Remember that when you look at the fraction you are looking at the relationship of the cross product to either the first vertical product, or the second vertical product.

Exercise 1:

	Base		Left		Right	Result	Answer
1.	6x6	2	-4	-4	16	2/16	
2.	6x7	3	-4	-3	12	3/12	
3.	6x8	4	-4	-2	08	4/08	
4.	6x9	5	-4	-1	04	5/04	
5.	7x7	4	-3	-3	09	4/09	
6.	7x8	5	-3	-2	06	5/06	
7.	7x9	6	-3	-1	03	6/03	
8.	8x8	6	-2	-2	04	6/04	
9.	8x9	7	-2	-1	02	7/02	
10.	9x9	1	-1	-1	01	8/01	
11.	6x12	8	-4	2	-8	8/-08	
12.	6x13	9	-4	3	-12	9/-12.	
13.	6x14	10	-4	4	-16	10/-16	
14.	6x15	11	-4	5	-20	11/-20	
15.	7x16	13	-3	6	-18	13/-18	
16.	7x17	14	-3	7	-21	14/-21	
17.	7x18	15	-3	8	-24	15/-24	
18.	7x19	16	-3	9	-27	16/-27	
19.	8x12	10	-2	2	-4	10/-04	
20.	8x13	11	-2	3	-6	11/-06	
21.	8x14	12	-2	4	-8	12/-08	
22.	8x15	13	-2	5	-10	13/-10	
23.	9x16	15	-1	6	-6	15/-06	
24.	9x17	16	-1	7	-7	16/-07	
25.	9x18	17	-1	8	-8	17/-08	
26.	9x19	18	-1	9	-9	18/-09	
27.	13*16	19	3	6	18	19/18	
28.	13*17	20	3	7	21	19/18	
29.	13*18	21	3	8	24	21/24	
30.	13*19	22	3	9	27	22/27	
31.	14*16	20	4	6	24	20/24	
32.	14*17	21	4	7	28	21/28	
33.	14*18	22	4	8	32	22/32	
34.	14*19	23	4	9	36	23/36	
35.	15*16	21	5	6	30	21/30	
36.	15*17	22	5	7	35	22/35	
37.	15*18	23	5	8	40	23/40	

38.	15*19	24	5	9	45	24/45
39.	16*16	22	6	6	36	22/36
40.	16*17	23	6	7	42	23/42
41.	16*18	24	6	8	48	24/48
42.	16*19	25	6	9	54	25/54
43.	17*17	24	7	7	49	24/49
44.	17*18	25	7	8	56	25/56
45.	17*19	26	7	9	63	26/63
46.	18*18	26	8	8	64	26/64
47.	18*19	27	8	9	72	27/72

3. Bi-directionality

One of the great things about mental math is that the calculations can generally be done from either direction, and a certain symmetry generally is present. What is so great about this is that the methods, shortcuts and techniques you have at your disposal then are twice as useful or powerful.

Let's consider for a moment some of the exercises we did in the previous chapter. One of them, for example, was 16x16. In this one. Each number was six units over a base, so we headed 6 to 16 to get 22, and linked that to 6*6= 36. The result of the calculation was 22/36 = 256. If we were to turn that number around, we would be calculating 61x61. Could we use the same method? The answer is yes! . Now we are saying that each number is 60 units over a base of one. From there we had 60 and 60. We could even use our fingers on this one too, by letting each finger represent 10 units, and your thumb representing 50. Now we just add the two sixes together to get 12, and multiply that answer by 10 to get 120. Then we simply tack 36/12/01 together to get

3721. The major point about this being that anything ending in one is fair game for the same technique we used for something that started with one. It really is no more difficult to multiply 16*16. than it is to multiply 61*61.

Let's try another calculation based upon the same thought process. You want to multiply 91x61 . You could do it by putting 90 on the fingers of one hand and 60 on the fingers of the other hand. First multiply them to get one link, and then add them to get the second link. The last link is simply 01. The links together would be 54/15/01= 5551. That calculation was about as simple was any calculation you ever do. It's a nice ,simple, straight left to right calculation. We can even apply this to three digit numbers that end in one. For example, if you wanted to multiply 121 times 141 you could easily do it using this method is long as you knew that 12x14 was 168. The components would be 168/26/01. The answer would be 17,061.

You could even do a two digit by three digit multiplication this way. If you wanted to multiply 81x121 for example . As long as you knew that 8x12 was 96 this should be very easy. Your components would be 96/20/01. Your answer would be 9801. Now let's do some exercises that use the same principle.

Exercise 2

	base	sum	links	answer
48.	11x21	02	03	02/03/01	231
49.	11x31	03	04	03/04/01	0341
50.	21x41	08	06	08/06/01	0861
51.	31x51	15	08	15/08/01	1581
52.	41x61	24	10	24/10/01	2501
53.	51x61	30	11	30/11/01	3211
54.	51x71	35	12	35/12/01	3621

| 55. | 61x71 | 42 | 13 | 42/13/01 | 4331 |
| 56. | 71x81 | 56 | 15 | 56/15/01 | 5751 |

    In other cases it may be better to look at incremental amounts over or under a base. Personally anything with a common denominator of one whether read from left to right, or right to left is very hard to resist as a fraction. As long as you realize that 01*01= 01 and skip actually actually calculating it out.

    If you think about it this is similar to adding fractions that have a common denominator of one. We did the same calculation in those cases by multiplying the leading digits, adding the fraction to get a whole number, and then squaring 1*1 to get the last digit of the link. Only now the story is different, we are talking about being over a base of one as opposed to having a fraction that has a denominator of one and is therefore a whole number.

    Essentially we are describing the same operation with two different stories. Which one you prefer is up to you, the big deal is to get the math right. The same applies in the other direction. If you look at 16*16 as two fractions you can work from left to right and get the links 01/12/36= 256 also.

    It's not that one method is better than the other at all. It is that in some cases it may be better to look at fractions and ratios. In other cases it may work better to think in terms of a base.

    The narrative can even be extended to a single digit number like nine and two digit numbers that end in a base of one. The method in that case is far more cumbersome than just multiplying and linking; so I will save both of us a lot of superfluous energy and skip the explanation of that. I hope you're getting the sense that different narratives can be applied to the same operation. In some cases one narrative may be preferred over another. In other cases there may not be a clear advantage to one narrative over the other. Have the right attitude and try to pick the tool that does the job best for you. That is what a craftsman would do. A

craftsman wouldn't try to use the same tool for everything.

4. Over/Under 100

We have also seen that a base of one can be used to multiply. We can take a number over a base of 10 and reverse it and multiply easily because of bi-directionality. Now we will use the same process for numbers that are close to a base of 100.

If we wanted to multiply 99 x99, we have seen how you can multiply numbers that are over or it as follows: 99 is one unit less than 100. Let's put one unit on one finger of the left hand, and one unit on one finger of the right hand. Now, since we are dealing with a deficit from 100 . Subtract one of those two numbers from 99 to get 98. Now let's multiply those two ones together to get +01. Here are the numbers we are going to link together: 98/01. Believe it or not 9,801 is our answer.   If   we were going to multiply 101x101 the narrative would be the same. He would have one on one finger of one hand and one on the finger of the other hand. This time since we are over a base . We add one to 101 to get 102. now we just tack 1x1 equals one onto 102 to get 10201.

If we were going to multiply 96x99 we would put four fingers out on our left hand, and one finger out on our right hand. We would then either subtract four from 99 to get 95; or subtract one from 96 to get 95. That 95 will be the first part of our answer. The second part of our answer will be the product of the numbers on our fingers. That would be 4x1 equals four. Our complete answer. Then would be 95/04= 9,504.

If we're going to multiply 102x104 we would put two fingers out on our left hand and four fingers out on our right hand. Then we would either add four to 102; or we could add two to 104. Either way, we get 106 as the first part of our answer. Then we multiply together. Then we multiply together the numbers on our hands. That would be 2x4= 08. Our complete answer then would be 10,608.

If we were going to multiply 98x101. We could put two fingers out on our left-hand, and one finger out on our right hand. Now either <u>subtract</u> two from 101, or add 1 to 98. Either way we get 99, and that is the first part of our answer. The second part is the product of the numbers on our fingers. One number is negative, the other one is positive. The product of the two numbers is -2x1= -2. In this case we subtract 2 from 9900 to get 9,898.

This is one of those processes that is so simple that if you have done it a few times , it becomes quite easy. The best thing to do at this point would be play with the numbers and see how they work together. Accordingly, here are some practice problems.

Exercise 3

	Problem:	Left	Right	1st part	2nd part	answer
1.	97x97	-03	-03	94	09	9409
2.	96x96	-04	-04	92	16	9216
3.	95x95	-05	-05	90	25	9025
4.	94x94	-06	-06	88	36	8836
5.	93x93	-07	-07	86	49	8649
6.	92x92	-08	-08	84	64	8464
7.	91x91	-09	-09	82	81	8281
8.	89x89	-11	-11	78	121	7921
9.	88x88	-12	-12	76	144	7744
10.	87x87	-13	-13	74	169	7569
11.	86x86	-14	-14	72	196	7396

12.	85x85	-15 -15	70	225	7225
13.	101x101	01 01	102	01	10201
14.	102x102	02 02	104	04	10404
15.	103x103	03 03	106	09	10609
16.	104x104	04 04	108	16	10816
17.	105x105	05 05	110	25	11025
18.	106x106	06 06	112	36	11236
19.	107x107	07 07	114	49	11449
20.	108x108	08 08	116	64	11664
21.	109x109	09 09	118	81	11881
22.	111x111	11 11	122	121	12321
23.	112x112	12 12	124	144	12544
24.	113x113	13 13	126	169	12769
25.	114x114	14	128	196	12996
26.	115x115	15	130	225	13225
27.	102x98	02	100	-04	9996
28.	102x104	02	106	08	10608
29.	98x96	-02	-04	08	9408
30.	85x99	-15	-01	15	8415
31.	85x115	-15	15	-225	9775
32.	98x114	-02	14	-28	11172
33.	91x113	-09	13	-117	10283
34.	89x121	-11	21	-231	10769

5. Over/Under 20,30...

I bet you're thinking to yourself right now that working over or under a base is very limited proposition. You're limited to numbers that are close to 1,10, or 100. Actually it is not true. With one single adjustment you can work with bases other than 1,10 ,or 100.

For example, if you wanted to multiply 23x31 we could do it by noting that 23 is seven less than a base of 30, and 31 is one more than a base of 30. now we have an established base, but it's not 10. Let's proceed anyway. Let's put minus seven on the fingers of the left hand, and 100 finger of the right hand. If we subtract seven from 31, or add 1 to 23 either way we get 24. This is where the other additional step comes in. We simply multiply that 24 by three because our base is 30. 24x3 equals 72, so that is our lead number. No multiplied together. The minus seven and the one to get -7. That is our second number. Now let's put it together 72/-07= 713. Notice how particularly easy this calculation is if you're dealing with a number that is only one unit over a base, as was 31 in this calculation. It is interesting that this problem slightly resembles a synchronized ratio or open carry problem in that the difference between the tens units is one.

If the numbers have the same base the calculation is still similar but without subtraction, and negative numbers. For example, if your multiplying 31x39. You could use 30 is a base too. Now you put one on the finger of your left hand, and nine on the finger of the right hand. If you add 9 to 31 you get 40, or if you add 1 to 39 you get 40. Now he multiply that 40 x 3 to get 120 and that is your lead number. The back number is the product of one and nine. Our answer is 1209.

If both numbers are closer to but under a base the calculation can be done in a way to assure the smallest possible adjustment to get your final answer.. For example, if you wanted to multiply 39x37. You could make 40 your base. Put -1 on the finger of your left hand. Put -3 on the fingers of your right hand. Take -1 from 37 to get 36. Now multiply that by four. To get an answer of 144. That is your leading number. The trailing number is -1x-3= +03. Your final answer then is 1443.

There is another way to solve this one it is an averaging problem also. You could square 38 to get 1444 and subtract one squared from that to get 1443.

This method would also work if you had to multiply numbers that were close a shared base such as 200 300, or 900 for that matter. 991X889 could be done by using a base of 900. One number is 91 units over 900, the other number is 11 units <u>under</u> 900. Subtract 11 from 991 to get 980. Now we multiply that by 9 to get 81/72 = 8820. That will be our lead number. -11x91= -1001. Our final answer is 8820/-1001= 880999. What looked to be a tough calculation wasn't so bad after all.

If the question had been to find the product of 911*989 the work would have been easier your lead number would be ((989+11)x9))= 9000. now we add to this 11x89= 979. Now the answer becomes 900979. In the exercises we will confine ourselves to two digit numbers with a base other than 10.

Vedic Math

Exercise 4

	lead	trail	answer
57. 21x39	90	-81	819
58. 24x36	90	-36	864
59. 27x33	90	-09	891
60. 87x93	81	-09	8091
61. 54x66	36	-36	3564
62. 73x87	64	-49	6351
63. 71x89	64	-81	6319
64. 82x98	81	-64	8036
65. 52x68	36	-64	3536

Have you noted anything interesting about these numbers? I deliberately picked the first nine because they can be looked at several different ways. In my previous books I called the first ten examples synchronized ratios, or open carry situations. The cross product of the calculation works out to be the smaller number of the calculation. There it is right under your nose.

This happens whenever the units digit sums to 10, and the tens digits are one unit apart. From a ratio standpoint the first problem amounts to adding the two fractions 3/9= 1/3+ 2/1= 2_ The total is 7/3, and 7/3x9=21. The shortcut is just to look at the numbers, see the characteristics of the units column and tens columns and just know that the cross product is 21.

The problem is also a averaging problem where each units digit is the tens compliment of the other. The shared base in this example is 30. One number is nine more than that, 39. The other number is nine less than 30, or 21. The numbers on your fingers are equal so their product is the square of nine in this case. That would make the adjustment to 90 be -81. Your answer is becomes 819.

Call them synchronized ratios, open carry,

averaging, or over and under a shared base. They all work out the same way. The smaller number is the cross product. There is no need to reinvent the wheel every time you see this situation. The cross product is right there in the open. The toolbox metaphor applies: a craftsman picks the right tool for the job.

The next seven problems are nines compliments, and the next two are eights compliments. The last one is a seven compliment. Again, these were hand picked too. Here is a list of the cross products for the last ten problems compared to what the cross product would have been if the larger number were a little bigger, just big enough in fact to be a tens compliment of the smaller number:

Tens CP: What it was: Difference:

1. 21	19	-2
2. 24	22	-2
3. 27	25	-2
4. 87	79	-8
5. 54	49	-5
6. 73	66	-7
7. 71	64	-7
8. 82	66	-16
9. 52	42	-10
10. 52	37	-15

Notice the pattern? A one unit change in the units sum results in a corresponding change in the real cross product in the amount of the tens cross product times the change amount.

In other words if the total of the units column is nine instead of ten just take that lead number of the cross product and subtract that digit from the tens cross product. This is amounts to a simple adjustment. In the case of problem 11 you take 21 and subtract two from it and you have the right cross product for 21x38. The rest of the solution is linking together the product of the lead digits and the trailing digits. 3X2/21-2/8x1= 6/19/08= 798.

I digressed off topic quite a bit on this, we will come back to this and all kinds of clever ways to use these properties to a math advantage later on.

Did you notice how small the adjustment became when we picked the base closest to the actual number being multiplied? That certainly makes the work easier. That is in fact is a good general strategy, pick the nearest base. To be sure there are other subtle ones that work especially well for larger numbers.

Your base does not have to be a multiple of ten or 100. You could pick one that had a digit sum of nine, or one that was a multiple of 11 for example. While not practical for a two digit situation, it might make a lot of sense for a three digit situation.

6. More on Ratios

Previously we talked about multiplying two digit numbers that had a units column that added up to 10, and the tens column was one unit in difference. It turns out that these numbers are over and under a base by an equal amount. For example, if you were to multiply 39x21. You could say that 39 is nine units over a base of 30, and 21 is nine units less than a base of 30.

Because one number is over a base in the other under a base. It turns out that the units are tens compliments of each other, or in other words they had to 10. When this happens , the smaller number is in fact the cross product. In the above example 21 is the cross product, and it's right there under your nose. So you could take a series of problems, like 39x21, 38x22, 37x23, 36x24, 35x25, 34x26, 33x27, 32x28, and 31x29 and easily solve them. The cross product in each case would be the numbers 21 through 29. You could repeat this process with numbers like 99x81. The same pattern would hold. In fact, for 3 digit multiplication. If you wanted to multiply 624x576 . You could do it pretty easily because the cross product would be 576. Then you would have components to put together of 6x5 equals 30, 576, and 1824. The one line component string would be 30/57/1824= 359424. The only difficult part of the whole calculation would be multiplying 76x24.

What I think is even more important to this whole discussion is what happens when you're units digit does not add up to 10. Since we are dealing with tens, the adjustment you need to make to your cross product is very simple. What if you wanted to multiply the 38x 21? In this case your units add up to nine instead of 10 . So you are one digit less than a situation where 21 would be your cross product. In this case , you can adjust for being one unit off by subtracting 1/10 of the tens unit unit from 21, so you subtract 2 from 21 to get 19 now you have the correct cross product. If the problem had been to multiply 39x22 you could arrive at the cross product by simply taking 2 and adding it to 22 to get 24 as your cross product.

This proportion holds true no matter what your units total is. Just look at the total, and ask yourself is this over or under ten? If it's over you will be adding your correction. If your total is under 10 you will be subtracting. The amount of the adjustment is that over or under amount multiplied by the tens unit of the smaller number. If you had to solve 87x76 you could do it straight line as 56/97/42= 6612. The cross product being 76+(3x7)= 76+21= 97.

Aliquot or fractional parts can be used too. For example if you were to multiply 53x42. Note that the tens units are one unit apart, and the units digits add up to five which is half of ten. Now your adjustment is ½ of 40, or 20. Your 2 stays tacked onto that so the cross product is 22. Also if the units add to 15 the problem is again an aliquot parts adjustment 1.5x your lower tens number plus the lower units number added back in. For example, 39x26= 06/36/54=1164. The 36 being 1.5x20=30 +06=36.

The next logical question is what happens if the tens units are two units apart. Guess what? The same adjustment is applied, this time using the smaller units digit as the adjustment. For example, if you wanted to multiply 49x21. What would be your cross product be? We can go and reinvent the wheel again. If we did this the cross product would be (4x1)+(9x2)=4+18=22. The easier way would be to note that all the conditions are in place to do this calculation as an adjustment to a synchronized ratio. Just add the one in 21 back into 21 to get 22.

I hope you are starting to think some more about this. What if the variance from a perfect open carry was one unit each? Well, let's look at that scenario. Let's say you wanted to multiply 48x23. The tens units add up to 11, so that is one unit more than ten. The tens units are 2 digits apart and that is one digit more than we would have in a perfect open carry or synchronized ratio situation. The adjustment in this case is equal to the sum of the digits in the smaller number being added to the smaller number. In this case we add 5 to 23 to get 28. Let's check that using vertical and crosswise. (4x3)+(8x2)= 12+16=28.

If one number were over and the other number was under then the adjustment would be the difference between the digits. For example 46x23, the components being 08/23+1/18= 1058. This one is a little tricky because 2-3=-1 but our adjustment is positive because 3-2=+1. You can expand this adjustment to any combination of two digit numbers you want. Let's look at 56x78. Here we are 4 units over 10, and one unit more than the required one digit. Our correction to 56 would be what? It would be (+1x11)+ (3x5)=26. The cross product is 82. You can get it by adding 1x6 to 4x5, or by taking the sum of the digits times one and adding 3x5 to that. Either way you get 26 to add to 56 to get 82.

The single digit multiplication model of this form is instructional and may be a good place to develop some confidence in the method. Let's say you wanted multiply 73x9. The units digits add up to 12, when the tens digits are six units from the required one digit difference. The first vertical multiplication is zero since zero times seven is zero. The cross product is 6x9+9 equals 63. The final vertical calculation is 27. Now we had the components 00/54+9/27= 63/27= 657

I'll be the first to admit that this is pretty lame, but I want to make the point that the narrative can even be applied to single digit numbers. To do the full multiplication you just assemble three components, one which happens to be zero. The second component of the cross product in this case is always half done for you in that the first part of it is zero.

An example with 11 as the smaller number may also be instructional. Let's say you wanted to multiply 74x11 In this case the tens digit is five units greater than the one we look for. The units digits add up to five, so that is five units <u>less</u> than what we need. Our cross product is ½ of 10+ 01= 06., plus the original 5x1 for a total of 11. The shortcut on this is that anything with a digit sum of 11 and two digits multiplied by 11 has a cross product of 11. Try it with 83x11, 92x11, 29x11 ect.

Anything with two digits you multiply by 11 has a cross product equal to the sum of the digits. That is why they say to add the neighbor to multiply by 11. 96x11 then would have a cross product of 15. You could get it by subtracting 3 from11 to get 8. Then add back in 7x1=7. The total is 15. The alternative would be to subtract 7 from 3 to get 4 and add that to the 11 to get 15.

In any event, here are some exercises where I want you to compute the cross product only using the synchronized ratio and adjusting off of that.

Exercises: Problem   components

1.	48x21	08/?/08
2.	48x23	08/?/24
3.	41x34	12/?/04
4.	64x23	12/?/12
5.	72x43	28/?/06
6.	84x52	40/?/08
7.	96x38	27/?/48
8.	61x43	24/?/03
9.	84x76	56/?/24
10.	72x49	28/?/18

Answers

1. 20
2. 28
3. 19
4. 26
5. 29
6. 36
7. 90
8. 22
9. 76

ABOUT THE AUTHOR

John Carlin lives in Apple Valley, Minnesota. He enjoys the beauty, symmetry, and immutability of math, and hopes you will come to appreciate it as well. He is a graduate of the U.S. Merchant Marine Academy in Kings Point, N.Y., and has an MBA from the University of Minnesota. Besides math, John is a music lover. His tastes run from Classical, to Jazz, R&B, and Rock.

Visit my website at www.binomialblvd.com

www.ingramcontent.com/pod-product-compliance
Lightning Source LLC
Chambersburg PA
CBHW072045190526
45165CB00018B/1842